Marine Mammals training Book

Mastering the
Art
of
Marine Mammals Training

Introduction: Training Marine Mammals: A Comprehensive Guide

Marine mammals, with their intelligence, agility, and captivating behaviors, have long fascinated both scientists and the public alike. From the playful dolphins to the majestic whales, these remarkable creatures inhabit the world's oceans, captivating our imaginations and inspiring awe and wonder. In captivity, marine mammal training plays a crucial role in not only ensuring their well-being but also providing opportunities for research, education, and conservation efforts.

"Training Marine Mammals: A Comprehensive Guide" is a definitive resource for anyone interested in understanding and mastering the art and science of training these extraordinary animals. Whether you're a seasoned trainer, a budding enthusiast, or someone simply intrigued by the world of marine mammals, this book offers valuable insights, practical techniques, and expert guidance to enhance your understanding and skills in marine mammal training.

Exploring the World of Marine Mammals

The book begins by delving into the fascinating world of marine mammals. From the Arctic waters to the tropical seas, we explore the diverse habitats, behaviors, and adaptations of whales, dolphins, seals, sea lions, and more. Understanding the natural history and biology of these animals provides a crucial foundation for effective training techniques.

Principles and Techniques of Animal Training

Building upon this understanding, the book explores the principles and techniques of animal training.

From the basics of operant conditioning to advanced training methodologies, readers will learn how to apply positive reinforcement, shaping, and other techniques to teach a wide range of behaviors to marine mammals.

Setting Up Successful Training Programs

Effective training programs require careful planning and clear objectives. This book provides practical guidance on setting training goals, developing training plans, and creating enriching environments for marine mammals. Whether training for public presentations, research tasks, or medical procedures, readers will learn how to design and implement successful training programs tailored to the needs of the animals.

Addressing Challenges and Ensuring Welfare

Training marine mammals is not without its challenges. From behavioral issues to safety concerns, this book offers strategies for addressing challenges and ensuring the welfare of both the animals and the trainers. Readers will learn how to identify and overcome behavioral challenges, implement safety protocols, and promote positive relationships with the animals.

Applications in Conservation and Education

Finally, the book explores the broader applications of marine mammal training in conservation and education. From using trained animals in research projects to engaging the public in educational programs, marine mammal training plays a vital role in raising awareness and fostering conservation efforts.

Conclusion

"Training Marine Mammals: A Comprehensive Guide" is a valuable resource for trainers, researchers, educators, and anyone passionate about marine mammals. With its blend of scientific insights, practical techniques, and real-world examples, this book equips readers with the knowledge and skills to build meaningful relationships with marine mammals and contribute to their well-being and conservation. Whether you're embarking on a career in marine mammal training or simply seeking a deeper appreciation for these magnificent animals, this book is your guide to unlocking the secrets of the sea's most captivating creatures.

Contants of the Book:

1. Understanding Marine Mammals
- Introduction to the diverse world of marine mammals
- Overview of different species, habitats, and behaviors
- Importance of understanding species-specific characteristics for effective training

2. Principles of Animal Training
- Learning theories and their application to marine mammals
- Positive reinforcement techniques
- Understanding operant conditioning and classical conditioning in training marine mammals

3. Setting Up a Successful Training Program
- Designing training goals and objectives
- Developing a training plan
- Creating an enriching environment for marine mammals

4. Building Trust and Relationship
- Importance of trust in animal training
- Establishing a strong bond with marine mammals
- Building mutual respect between trainers and animals

5. Basic Training Techniques
- Target training
- Desensitization and counterconditioning
- Shaping behaviors through successive approximations

6. Advanced Training Techniques
- Complex behavior chains
- Stimulus discrimination
- Generalization of behaviors in different contexts

7. Training for Husbandry and Medical Procedures
- Training for voluntary medical examinations
- Desensitization to medical equipment
- Encouraging cooperative behaviors for health care
-

8. Addressing Behavioral Challenges
- Dealing with aggression and fear
- Problem-solving in training sessions
- Modifying undesirable behaviors
-

9. Training for Public Presentation and Education
- Designing engaging presentations
- Teaching natural behaviors for educational purposes
- Ensuring the welfare of animals in public settings

10. Ethical Considerations and Future Directions
- Ethical guidelines in training marine mammals
- Balancing conservation, education, and entertainment
- The future of marine mammal training: innovations and challenges

Chapter 1: Understanding Marine Mammals

Introduction

Marine mammals, an extraordinary group of animals encompassing whales, dolphins, seals, sea lions, manatees, and others, have captivated human interest for centuries. From the majestic humpback whale breaching the ocean's surface to the playful antics of dolphins riding waves, these creatures inhabit a diverse range of habitats, from the icy waters of the Arctic to the tropical seas of the equator. Understanding marine mammals is not only fascinating but also crucial for their conservation and welfare, especially in the context of training.

The Diversity of Marine Mammals

Marine mammals represent a remarkable diversity of species, each adapted to its unique aquatic environment. Whales, the largest marine mammals, include the immense blue whale, the acrobatic humpback, and the highly intelligent sperm whale. Dolphins, known for their social behaviors and intelligence, inhabit oceans worldwide, from coastal waters to the open sea. Seals and sea lions, with their sleek bodies and flipper-like limbs, are well-adapted to both land and water, while manatees and dugongs are the gentle giants of tropical waters, grazing on underwater vegetation.

Habitats and Behaviors

Understanding the habitats and behaviors of marine mammals is essential for their management and training. Whales undertake epic migrations, traveling thousands of miles annually between feeding and breeding grounds. Dolphins exhibit complex social structures, often living in large groups called pods, while seals and sea lions gather in colonies on land for breeding and molting. These animals display a wide range of behaviors, from breaching and spy-hopping to echolocation and vocal communication.

Adaptations to Aquatic Life

Marine mammals have evolved a suite of adaptations to thrive in their underwater world. Their streamlined bodies reduce drag, enabling swift movement through the water, while their thick blubber provides insulation against cold temperatures. Specialized senses, such as echolocation in toothed whales and sensitive whiskers in seals, help them navigate and find food in the ocean's depths. Marine mammals also possess remarkable physiological adaptations, such as the ability to hold their breath for extended periods and regulate their body temperature in cold waters.

Conservation Challenges

Despite their remarkable adaptations, marine mammals face numerous threats in the modern world. Habitat degradation, pollution, climate change, entanglement in fishing gear, and collision with vessels are significant challenges for their survival. Overfishing and depletion of prey species also impact marine mammal populations, while noise pollution from human activities disrupts their communication and behavior. Understanding these threats is crucial for developing effective conservation strategies.

The Role of Training

Understanding the natural history and biology of marine mammals is fundamental to their successful training. By appreciating their natural behaviors, trainers can develop training programs that are enriching, safe, and respectful of the animals' welfare. Training also plays a vital role in conservation efforts, enabling researchers and wildlife managers to study and monitor marine mammal populations, as well as in public education and outreach programs aimed at raising awareness about these remarkable animals.

Conclusion

Understanding marine mammals is a multifaceted endeavor that encompasses their biology, behavior, and ecology. From the icy waters of the Arctic to the coral reefs of the tropics, these animals continue to inspire awe and wonder. By studying and appreciating the diversity of marine mammals, we can work towards their conservation and welfare, ensuring that these magnificent creatures thrive in our oceans for generations to come.

Chapter 2: Principles of Animal Training

Introduction

Effective training of marine mammals relies on a solid understanding of the principles of animal learning and behavior. By applying these principles, trainers can foster positive relationships with the animals, facilitate learning, and achieve desired behavioral outcomes. This chapter explores key concepts in animal training and their application to marine mammals.

Learning Theories

Understanding the various learning theories is fundamental to successful animal training. Two main theories are particularly relevant:

1. Operant Conditioning:
Developed by psychologist B.F. Skinner, operant conditioning focuses on the relationship between behavior and its consequences. Behaviors that are reinforced tend to be repeated, while those that are not reinforced or are punished are less likely to occur. Marine mammal trainers use operant conditioning techniques to shape desired behaviors through reinforcement, such as food rewards or positive social interactions.

2. Classical Conditioning:
Classical conditioning involves associating a neutral stimulus with a naturally occurring stimulus to elicit a specific response. Ivan Pavlov's experiments with dogs famously demonstrated this concept. In marine mammal training, trainers use classical conditioning to create positive associations between a neutral

stimulus (like a whistle or hand signal) and a desired behavior, enabling the animals to respond predictably to cues.

Positive Reinforcement Techniques

Positive reinforcement is a cornerstone of effective marine mammal training. By providing a desirable consequence (such as food, toys, or praise) immediately following a desired behavior, trainers increase the likelihood of that behavior recurring in the future. Key points about positive reinforcement include:

- Timing: Reinforcers must be delivered immediately after the desired behavior to be effective.
- Consistency: Reinforcement should be consistent to reinforce the desired behavior.
- Variety: Using a variety of reinforcers keeps training sessions engaging for the animals.

Understanding Operant Conditioning

Operant conditioning involves four main components:

1. Positive Reinforcement (R+): Adding a desirable stimulus to increase the likelihood of a behavior occurring again.
2. Negative Reinforcement (R-): Removing an aversive stimulus to increase the likelihood of a behavior recurring.
3. Positive Punishment (P+): Adding an aversive stimulus to decrease the likelihood of a behavior occurring again.
4. Negative Punishment (P-): Removing a desirable

stimulus to decrease the likelihood of a behavior recurring.

Marine mammal trainers primarily focus on positive reinforcement (R+), using rewards such as fish, toys, or tactile interactions to reinforce desired behaviors.

Training Techniques

Several techniques are used in marine mammal training to shape behaviors effectively:

- Target Training: Teaching the animal to touch a specific object, such as a target pole, with a body part, usually the nose or flipper.
- Shaping: Breaking down complex behaviors into small, manageable steps and reinforcing each step until the desired behavior is achieved.
- Capturing: Reinforcing spontaneous occurrences of the desired behavior.

Conclusion

By applying principles of animal learning and behavior, trainers can establish effective and respectful relationships with marine mammals. Positive reinforcement techniques, coupled with an understanding of operant conditioning, allow trainers to shape behaviors, facilitate learning, and enhance the welfare of the animals under their care. In the following chapters, we will explore how these principles are applied in the development of successful training programs for marine mammals.

Chapter 3: Setting Up a Successful Training Program

Establishing a successful training program for marine mammals requires careful planning, clear objectives, and a deep understanding of the animals' needs and behaviors. This chapter outlines the essential steps in setting up such a program, from defining goals to creating an enriching environment for the animals.

Defining Training Goals and Objectives

Before beginning a training program, it's crucial to define clear goals and objectives. These may include:

- Behavioral Goals: Desired behaviors to be trained, such as high jumps, vocalizations, or participating in research tasks.
- Health and Husbandry Goals: Behaviors related to medical procedures, such as voluntary blood draws or body examinations.

Enrichment Goals: Providing mental and physical stimulation to ensure the animals' well-being.

Developing a Training Plan

Once goals and objectives are established, a comprehensive training plan should be developed. This plan outlines:

- Training Techniques: Methods to be used for teaching each behavior, such as shaping, capturing, or targeting.
- Reinforcement Strategies: The types of rewards to be used, including food, toys, or social interactions.
- Progression: A step-by-step approach for shaping behaviors, from initial approximations to the final desired behavior.

- Criteria for Success: Clear criteria for when a behavior is considered learned and ready to be reinforced.
- Evaluation: Methods for assessing progress and adjusting the training plan as needed.

Creating an Enriching Environment

An enriching environment is essential for the physical and psychological well-being of marine mammals. Elements of an enriching environment include:

- Physical Enrichment: Providing a variety of toys, objects, and structures for the animals to interact with, such as balls, ropes, and floating platforms.
- Social Enrichment: Facilitating social interactions between animals, either within their own species or with other compatible species.
- Cognitive Enrichment: Offering opportunities for mental stimulation through problem-solving tasks, puzzles, and training sessions.
- Sensory Enrichment: Stimulating the animals' senses through auditory, visual, and olfactory stimuli, such as music, mirrors, and scents.

Implementing the Training Program

Once the training plan is in place, it's time to implement it with the animals. Key points to consider include:

- Consistency: Maintaining consistency in training sessions, including timing, cues, and reinforcement.
- Patience and Persistence: Understanding that learning takes time and that progress may be gradual.
- Flexibility: Being open to adjusting the training plan based on the animals' responses and progress.

- Safety: Ensuring the safety of both the animals and the trainers throughout the training process.

Monitoring and Evaluation

Regular monitoring and evaluation are essential to track progress and make adjustments as needed. This may involve:

- Recording Progress: Keeping detailed records of training sessions, including behaviors trained, progress made, and any challenges encountered.
- Observation: Monitoring the animals' responses during training sessions to assess their understanding and comfort level.
- Feedback and Communication: Maintaining open communication among trainers and with animal care staff to share insights, feedback, and observations.

Conclusion

A successful training program for marine mammals is built on a foundation of clear goals, effective planning, and a stimulating environment. By defining training objectives, developing a comprehensive plan, and implementing it with patience and consistency, trainers can foster positive relationships with the animals and achieve desired behavioral outcomes. In the following chapters, we will delve deeper into specific training techniques and their applications in various contexts.

Chapter 4: Building Trust and Relationship

Introduction

Establishing a strong bond based on trust and mutual respect is fundamental in training marine mammals. This chapter explores the importance of building trust and relationships with these animals, along with practical techniques to achieve this.

Understanding Trust in Animal Training

Trust is the cornerstone of any successful animal training program. When marine mammals trust their trainers, they are more likely to engage in training sessions willingly and exhibit the desired behaviors. Trust is built through:

- Consistency: Consistent interactions and training sessions help animals feel secure and confident.
- Positive Reinforcement: Using positive reinforcement techniques fosters trust by associating training with pleasant experiences.
- Respect: Respecting the animal's boundaries and preferences reinforces trust and builds a positive relationship.
- Clear Communication: Using clear, consistent signals and cues helps the animal understand what is expected, reducing confusion and stress.

Establishing a Connection

Building a strong connection with marine mammals involves several key elements:

- Spending Time Together: Spending quality time with the animals outside of training sessions helps strengthen the bond.

- Observation and Understanding: Observing the animal's natural behaviors and understanding its communication signals are essential for building rapport.
- Positive Associations: Creating positive associations with the trainer and training environment through enjoyable activities and interactions.

Effective Communication

Clear communication is vital for building trust and facilitating learning:

- Body Language: Marine mammals are highly attuned to body language. Using calm, relaxed body language conveys trust and reduces stress.
- Voice Tone: Using a calm, reassuring tone of voice helps the animal feel safe and secure.
- Consistent Signals: Consistency in signals and cues helps the animal understand what is expected and builds confidence.

Building Confidence

Building confidence in marine mammals is crucial for their willingness to engage in training:

- Successive Approximations: Breaking down behaviors into small steps and reinforcing each step helps build confidence as the animal experiences success.
- Positive Feedback: Providing positive feedback, such as praise or rewards, reinforces the animal's confidence in its abilities.
- Challenging but Achievable Goals: Setting goals that are challenging yet achievable encourages the animal to stretch its abilities and build confidence.

Respect for Individual Differences

Each marine mammal is unique, with its own personality, preferences, and learning style. Respecting these individual differences is essential for building trust and rapport:

- Tailoring Training Approaches: Adapting training techniques to suit the individual animal's temperament and learning style fosters trust and cooperation.
- Building on Strengths: Identifying and building on the animal's strengths helps boost confidence and motivation.

Conclusion

Building trust and relationships with marine mammals is a gradual process that requires patience, consistency, and respect. By establishing clear communication, spending quality time together, and building confidence through positive reinforcement, trainers can develop strong bonds with these remarkable animals. These bonds not only enhance the training process but also contribute to the overall well-being and welfare of the animals under their care. In the following chapters, we will explore specific training techniques to further develop these relationships and achieve training goals.

Chapter 5: Basic Training Techniques

Introduction

Basic training techniques are the foundation of any successful marine mammal training program. In this chapter, we will explore the fundamental techniques used to teach and reinforce behaviors in marine mammals, laying the groundwork for more advanced training.

Target Training

Target training is a versatile technique used to teach marine mammals to touch a specific object, usually a buoy or pole, with a particular body part, typically the nose or flipper. The steps for target training include:

1. Introduction to the Target: Present the target to the animal and wait for any interaction.
2. Shaping the Behavior: Gradually reinforce the animal for getting closer to the target until it touches it with the desired body part.
3. Reinforcement: Reward the animal immediately upon touching the target.
4. Increasing Complexity: Once the animal reliably touches the target, begin introducing variations such as distance, duration, and different locations.

Positive Reinforcement

Positive reinforcement is a powerful tool for shaping behaviors in marine mammals. It involves providing a desirable consequence, such as food, toys, or praise, immediately following a desired behavior. Key points about positive reinforcement include:

- Timeliness: Reinforcement must be given immediately after the desired behavior to be effective.
- Consistency: Reinforcement should be consistent to reinforce the desired behavior.
- Variety: Using a variety of reinforcers keeps training sessions engaging for the animals.

Shaping Behaviors

Shaping is the process of breaking down complex behaviors into smaller, manageable steps and reinforcing each step until the desired behavior is achieved. The steps for shaping behaviors include:

1. Identifying the Target Behavior: Clearly define the behavior you want to teach.
2. Breaking it Down: Break the behavior into small, achievable steps.
3. Reinforcing Each Step: Reinforce each approximation of the behavior until the final behavior is achieved.
4. Fading Prompts: Gradually reduce any prompts or aids used to elicit the behavior.

Capturing Behaviors

Capturing involves capturing spontaneous occurrences of the desired behavior and reinforcing them. The steps for capturing behaviors include:

1. Observation: Watch the animal closely for any occurrences of the behavior you want to capture.
2. Timing: Immediately reinforce the behavior when it occurs.
3. Consistency: Be consistent in reinforcing the behavior to strengthen its occurrence.

Using Bridges and Secondary Reinforcers

Bridges, also known as conditioned reinforcers, are signals that indicate to the animal that a correct behavior has been performed and that a reward is forthcoming. Secondary reinforcers, such as a whistle or clicker, are commonly used as bridges. The steps for using bridges and secondary reinforcers include:

1. Pairing the Bridge with the Primary Reinforcer: Associate the bridge with the delivery of a primary reinforcer, such as food.
2. Timing: Use the bridge immediately after the correct behavior occurs.
3. Consistency: Be consistent in using the bridge to signal the correct behavior.

Conclusion

Basic training techniques provide the groundwork for teaching a wide range of behaviors to marine mammals. By using target training, positive reinforcement, shaping, capturing, and bridges, trainers can effectively communicate with the animals and facilitate learning. In the following chapters, we will delve into more advanced training techniques and their applications in different contexts.

Chapter 6: Advanced Training Techniques

Introduction

Advanced training techniques build upon the foundation established by basic training methods, allowing trainers to teach more complex behaviors and refine existing ones. In this chapter, we will explore several advanced techniques used in marine mammal training.

Complex Behavior Chains

Behavior chains involve linking together multiple individual behaviors to create a sequence. Complex behavior chains consist of several individual behaviors performed in a specific order to achieve a final outcome. The steps for training complex behavior chains include:

1. Breaking Down the Chain: Identify each individual behavior within the chain and teach them separately.
2. Linking Behaviors: Once each behavior is mastered, begin linking them together into a chain.
3. Reinforcing Each Step: Reinforce each behavior within the chain to strengthen the sequence.
4. Maintaining Fluency: Practice the behavior chain regularly to maintain fluency and accuracy.

Stimulus Discrimination

Stimulus discrimination involves teaching an animal to differentiate between different cues or stimuli and respond appropriately to each. The steps for training stimulus discrimination include:

1. Introducing Discriminative Stimuli: Present two or more stimuli and reinforce the correct response to each.
2. Gradually Differentiating Stimuli: Increase the difference between the stimuli until the animal reliably responds only to the target stimulus.
3. Generalizing Discrimination: Practice the discrimination in different contexts to ensure the animal responds correctly in various situations.

Generalization of Behaviors

Generalization involves teaching an animal to perform a behavior in a variety of contexts and environments. The steps for training generalization include:

1. Varying Conditions: Practice the behavior in different locations, with different trainers, and amidst distractions.
2. Changing Stimuli**: Introduce variations in cues, objects, or conditions while maintaining the integrity of the behavior.
3. Reinforcing Generalization: Reinforce the behavior in each new context to strengthen its generalization.

Backchaining

Backchaining involves teaching a behavior chain by starting with the last behavior and working backward. The steps for training a behavior chain using backchaining include:

1. Identifying the Final Behavior: Determine the last behavior in the chain.
2. Teaching the Last Behavior: Teach and reinforce the last behavior until it is mastered.

3. Adding the Previous Behavior: Once the last behavior is mastered, add the second-to-last behavior to the chain.
4. Continuing Backward: Continue adding behaviors to the chain in reverse order until the entire sequence is completed.

Shaping More Complex Behaviors

As behaviors become more complex, shaping techniques may need to be refined. This can include:

- Chaining: Linking together multiple behaviors into a sequence.
- Fine-Tuning: Making subtle adjustments to behavior criteria to achieve precision.
- Adding Variability: Introducing variations within a behavior to increase flexibility.

Conclusion

Advanced training techniques allow trainers to teach complex behaviors, promote discrimination and generalization, and refine existing skills in marine mammals. By incorporating techniques such as behavior chaining, stimulus discrimination, and backchaining, trainers can enhance the animals' cognitive abilities and expand their repertoire of behaviors. In the following chapters, we will explore how these advanced techniques are applied in specific training contexts, such as medical procedures, public presentations, and research tasks.

Chapter 7: Training for Husbandry and Medical Procedures

Introduction

Training marine mammals for husbandry and medical procedures is essential for their health and well-being in captivity. By teaching these animals to participate voluntarily in their own care, trainers can reduce stress, minimize the need for restraint, and facilitate necessary medical interventions. In this chapter, we will explore the techniques and strategies used to train marine mammals for various husbandry and medical procedures.

Understanding the Importance of Husbandry Training

Husbandry training involves teaching marine mammals to participate voluntarily in behaviors related to their daily care and health management. These behaviors may include:

- Body Presentations: Allowing trainers to observe the animal's body for signs of injury, illness, or abnormalities.
- Body Movements: Assisting with routine movements such as positioning for medical exams or transport.
- Medical Procedures: Participating in medical exams, treatments, and diagnostic procedures.

Positive Reinforcement in Husbandry Training

Positive reinforcement techniques are used extensively in husbandry training to encourage the animals to willingly participate in their care. By associating these behaviors with positive experiences and rewards, trainers can make the process enjoyable for the animals. Key points in positive reinforcement for husbandry training include:

- Gradual Introduction: Introduce husbandry behaviors gradually, starting with simple, non-invasive tasks.
- Consistent Reinforcement: Reinforce each desired behavior consistently to maintain motivation and cooperation.
- Building Trust: Use husbandry training as an opportunity to build trust and strengthen the bond between trainer and animal.

Training Specific Husbandry Behaviors

Voluntary Body Presentations: Teaching the animal to present different parts of its body for inspection allows trainers to monitor its health and well-being. Steps for training voluntary body presentations include:

1. Introduction to Touch: Get the animal comfortable with being touched and manipulated by the trainer.
2. Targeting Body Parts: Use target training to teach the animal to target different body parts to specific locations.
3. Progressive Desensitization: Gradually introduce equipment used in medical exams, such as stethoscopes or ultrasound probes.

Medical Procedures

Training marine mammals for medical procedures is crucial for both their health and the safety of veterinary staff. Common medical procedures trained include:

- Voluntary Blood Draws: Teaching the animal to present a specific body part for blood collection.
- Ultrasound Exams: Training the animal to hold still or present the desired body part during ultrasound examinations.
- Dental Exams and Care: Allowing the animal to open its mouth for dental inspections or tooth cleaning.
- Injections and Medications: Teaching the animal to accept injections or medications voluntarily.

Safety Considerations

Ensuring the safety of both the animals and the trainers during husbandry and medical procedures is paramount. Safety considerations include:

- Positive Association with Equipment: Introduce medical equipment gradually and associate it with positive experiences.
- Emergency Procedures: Have protocols in place for handling emergency situations during training or medical procedures.
- Trainer Awareness: Trainers should be aware of the animal's body language and signs of stress or discomfort.

Conclusion

Training marine mammals for husbandry and medical procedures is a crucial aspect of their care in captivity. By using positive reinforcement techniques and gradually introducing behaviors, trainers can teach the animals to participate voluntarily in their own care, promoting their health and well-being. In the next chapter, we will explore how training techniques are applied in the context of public presentations and educational programs.

Conclusion: Training Marine Mammals: A Comprehensive Guide

As we reach the conclusion of "Training Marine Mammals: A Comprehensive Guide," we reflect on the journey we've taken through the fascinating world of marine mammal training. From understanding the natural behaviors of whales, dolphins, seals, and sea lions to mastering the techniques of positive reinforcement and behavior shaping, this book has provided a thorough exploration of the art and science of training these remarkable animals.

Throughout the chapters, we've emphasized the importance of building trust, fostering positive relationships, and prioritizing the welfare of the animals. We've learned how to set clear training goals, develop effective training plans, and create enriching environments that promote learning and well-being.

We've explored the challenges that trainers may encounter, from addressing behavioral issues to ensuring safety during medical procedures. By employing strategies such as desensitization, behavior modification, and collaboration among trainers, we've seen how these challenges can be overcome, leading to a safer, more enriching training environment for both animals and trainers alike.

Furthermore, we've examined the broader applications of marine mammal training in conservation, research, and education. By engaging the public in educational programs and supporting conservation efforts, trained marine mammals play a vital role in raising awareness and inspiring action to protect our oceans and their inhabitants.

As we close this book, it's clear that the training

of marine mammals is not just a profession but a passion—a dedication to understanding, respecting, and connecting with these extraordinary creatures. Whether you're a seasoned trainer, a budding enthusiast, or simply someone who appreciates the beauty of marine life, we hope this guide has deepened your understanding and appreciation for marine mammals and the important role they play in our world.

May the knowledge and techniques shared in these pages continue to inspire curiosity, compassion, and stewardship for marine mammals and their ocean home. Let us carry forward the lessons learned here as we work together to ensure a bright future for these magnificent animals and the oceans they call home. Thank you for joining us on this journey.

www.ingramcontent.com/pod-product-compliance
Lightning Source LLC
Chambersburg PA
CBHW050254230526
45470CB00005B/2265